———■———

*I dedicate this book
to my beautiful children,
Jeshua and Illana.*

———■———

TABLE OF CONTENTS

Once Upon a Complex Time: Using Stories to Understand Systems

ONCE UPON A COMPLEX TIME: USING STORIES TO UNDERSTAND SYSTEMS

by Richard D. Brynteson

Sparrow Media Group
Farmington, MN

ONCE UPON A COMPLEX TIME: USING STORIES TO UNDERSTAND SYSTEMS

Publisher's Cataloging-in-Publication Data

Brynteson, Richard D.
        Once upon a complex time : using stories to understand systems / by Richard Brynteson. - - 1st ed. - - Eden Prairie, Minn. : Sparrow Media Group, Inc. 2006.

        80 p. : cm.
        Includes bibliographic references and index.
        1. Systems theory. 2. Organizational behavior I. Title

Book and cover design by Avallo   www.avallo.com
Printed and bound in the United States.
First edition, 2006.

LCCN: 2006925103
ISBN: 0-9719304-8-1

Quantity discounts are available for organizations or universities. Contact:

Sparrow Media Group

16588 Fieldcrest Avenue, Farmington, MN 55024
Phone: 952-953-9166 • Fax: 952-431-3461
info@sparrowmediagroup.com • www.sparrowmediagroup.com

Pushback, Blowback, and Unintended Consequences

High-Leverage Interventions

INTRODUCTION

I could never figure it out. If my parents really cared about me, they would quit smoking. They knew that it irritated us kids. They should just stop doing it. Just like that! Now that I have worked with systems thinking, I realize that they couldn't stop, just like that. There is the smoking system—the process of lighting up, inhaling, exhaling, and flicking the ashes (and, maybe, coughing). But other systems also surrounded smoking: social systems, body addiction systems, emotional systems, thought systems, memory systems, and drinking systems. And, then, there are social systems—all of their friends smoked, at a time it was cool, and it tasted so good. These other systems keep the smoking pattern in place as much as anything else.

Systems thinking is thinking with a wide-angle lens.

Systems thinking is thinking with a wide-angle lens, not a telephoto lens. Systems thinking is seeing the connections between parts, not just the parts themselves. Systems thinking is seeing the patterns and structures underneath events, not just the events themselves. Systems thinking is examining the time and distance between cause and effect. Systems thinking is circular, not linear, thinking. Systems thinking is an excellent problem-solving tool.

Systems theory emerged in the beginning of the twentieth century as an outgrowth of quantum theory. Scientists saw systems theory as a way to explain the natural world. While in its infancy, systems theory was relegated to the sciences. However, by the middle of the twentieth century, systems theory had spread to the social and behavioral sciences. Murray Bowen developed family systems theory as a foundation for family therapy. In the 1970s, Jay Forrester and Dona Meadows used systems theory to predict population growth patterns in their classic *Limits to Growth*. In 1990, Peter Senge and others brought systems theory into the organizational studies with the publication of *The Fifth Discipline*. These authors built on each other's work and others to create the field called systems theory or systems thinking.

A body of systems theory literature has emerged in organizations. In 1980, Draper Kauffman wrote *Systems One* to use as a primer for systems thinking. The book outlines and illuminates the principles of systems with easy-to-understand examples. He was followed by Kim *(Introduction to Systems Thinking)*, Haines *(The Complete Guide to Systems Thinking and Learning)*, and Senge *(The Fifth Discipline* and *The Fifth Discipline Fieldbook)*. These books strengthened this body of knowledge immensely. These works have generally followed a pattern of moving from the general principle to examples.

In this book, I reverse this methodology by moving from the specific example to the underlying principle. I use stories to illuminate systems theory.

Stories are powerful. From Jesus to Margaret Thatcher, successful leaders have used stories to captivate followers. Stories awaken and inspire to action; they touch people at a very deep level and resonate at a soul level.

These stories are drawn from business, government, nature, technology, and interpersonal relationships. They tell about good people trying to do good things but ending up, often, with deleterious side effects. The results of their actions are often the opposite of what they desired—not because they were dumb or malicious, but because they did not understand the complexities of the systems in which they were working.

These systems stories invite the reader to examine themselves and their actions on a deeper level and to understand the complexities of life. These stories invite us to look at our illusion of control and to laugh at ourselves and other people, as we stand in true humility. These stories invite us to begin to truly understand the world around us beyond ten-second sound bites—to use a wide-angle lens when viewing the world. These stories also invite us to look for patterns in the complex web of natural and organizational life.

Finally, enjoy these stories. They are interesting and fun. Stories are tiny windows through which to view the world.

Richard D. Brynteson

THE SYSTEM YOU SEE IS BIGGER THAN IT APPEARS

These stories illustrate that systems are complex and unpredictable and, often, have unintended behavior. When we deal with complex systems, we are typically dealing with much more than meets the eye.

- WAY TOO OLD
- QUIET SUZY
- SEND THEM AWAY
- THE KNEE IS CONNECTED TO THE SHIN BONE
- THE COLUMBIA DISASTER
- CLOSED OR OPEN

WAY TOO OLD

Several years ago, I was teaching a university level class in systems thinking. One evening, I walked into class with two black eyes. "What happened?" asked a sympathetic student. "I was playing goalie on my soccer team. I came out of the net to block a shot and took the puck right in the face and it broke my glasses. It was a freak occurrence."

One week later, I walked into the same class on crutches and plopped down on a chair. "What happened this time?" inquired a student. "I was trying to cut down the angle on a breakaway. I slid into the oncoming attacker and pulled my muscle. But I stopped the goal and was a hero! It was purely an accident."

One week after that, I walked into the same class with two broken fingers. The same question was posed. "I leapt into the air for a high shot and the ball caught the ends of my fingers, bending them backwards and breaking them. It was a freak accident. They shouldn't have broken."

"But, Richard, these are not just random events! What is the pattern here?" asked Sean, a rather astute student. He was right. I try to help my students see the world in terms of patterns and structures that support the patterns and not just in random, freak events. In this case, I was seeing these events as random events. In fact, the pattern was a series of injuries, one per game. The underlying structure that created the pattern: I was a 47-year-old playing in a 20-year-old's league. That structure created the pattern of injuries to me. I retired shortly after that insight.

Reflection. Is there anything in this story that rings true for you? Are disconcerting events at work home connected? Does your child's pattern of acting out have an underlying structure—perhaps it occurs after a heavy dose of sugar? Is an employee late or absent every Monday? The underlying structure for that pattern might be alcohol consumption over the weekend.

Try this. Your homework for the rest of your life is to be a systems sleuth! Move from an event orientation to an event-pattern-structure orientation. When something disconcerting happens, reflect on it to see if there were any other similar events that have also happened. What is the pattern? And if there is a pattern, what is the underlying structure causing the pattern?

———————■———————

QUIET SUZY

It happens all over the United States, from Topeka to Tucson to Tacoma and from Portland, New Hampshire to Portland, Maine. It happens in first grade and in twelfth grade and every grade in between.

Ms. Peterson, third-grade idealistic teacher, walks into her classroom every morning and is greeted, or not greeted, by 22, 28, or 34 children. If she is lucky, she has an assistant for half of the year, but often, she is alone. She teaches these children from 8:00 a.m. to 2:30 p.m. She might have lunchroom duty, playground duty, or parent meetings sprinkled throughout the day.

The tragedy of the commons occurs when individuals or groups usurp public domain resources at the expense of others within the system.

Unfortunately, she has three or four students who just do not receive sufficient attention at home from their parents. These children are left in front of video games or the TV for long periods of time. Thus, the children compensate by demanding more of Ms. Peterson's attention. They are darling children, but they require extra time and attention. They need continual help focusing and extra help in understanding worksheets. They need more of her energy.

Quiet Suzy does her homework on time and easily understands worksheets because she listens attentively. She is given sufficient time and attention at home, so she does not crave it from Ms. Peterson. Because Suzy is low maintenance, Ms. Peterson can spend her time on more pressing concerns. Inadvertently, she mostly ignores Suzy.

Reflection. This is a perfect example of the systems' archetype *Tragedy of the commons* (Hardin, 1968). Several centuries ago, land was held in common in England. Sheep owners saw this as an opportunity to expand their herds using the public land to graze. It cost them no more but could enrich them by enlarging their herds. The tragedy of the commons occurs when individuals or groups usurp public domain resources at the expense of others within the system.

Ms. Peterson's time is the commons and it is increasingly gobbled up by attention-seeking children. The parent who uses TV or electronic devices as their children's babysitter is dooming the Suzys of the world to a substandard education. The teacher's time is a common, limited resource to be shared between children. There are bound to be winners and losers in this zero sum game.

Try this. Observe when the tragedy of the commons plays itself out in your work or home life? Does one department take more of the information technology department's time than they should, leaving other departments in need? Does one member of your family steal more of the Ben and Jerry's ice cream, leaving other family members high and dry? The first method of becoming a better thinker is watching how these systems work out in real life.

———————■———————

SEND THEM AWAY

A local large hospital in rural Midwest was going through major problems. It had lost a large part of its business due to the events in and around the 9/11 attacks on the Twin Towers and the Pentagon. As a result, it had to examine how to cut its costs. One way the hospital cut its costs was to reject Medical Assistance low-income clients to its psychiatric center. The hospital sent these clients to a non-profit counseling agency.

This non-profit counseling agency accepted the clients. But because of low reimbursement rates for Medical Assistance clients, they lost about $190 per-client, per-hour for the psychiatric services. As a result of this influx of clients, this agency lost so much money that it endangered the viability of the rest of the agency.

The internal tension came from two systems clashing.

The agency was faced with a dilemma. Continuing to accept these patients jeopardized the other community services they provided. Yet, the agency's mission was to provide services to low-income clients. Therefore, they had to serve these clients even though they couldn't serve these clients as they wanted. What should they do?

Reflection. The ripple effects of 9/11 are hard to gauge. In retrospect, the chain effect from 9/11 to the financial problems of this agency is easy to follow but hard to predict beforehand. The end result was a systemic dilemma that the agency must face and solve. Now, if the agency stops servicing these clients, they would likely go to a public facility, which would create an additional burden on that facility and county budgets. The chain effect will continue. Should the agency live by its mission and, perhaps, die or change its mission? The poor, then, might not have access to these services.

Try this. Take a significant event such as the fall of the Iron Curtain in 1989, the Internet, 9/11, or the 2004 tsunami and trace the effect to organizations and people that you know. For instance, a gentleman named Mohammed, who spoke little English, searched me at length after a 20-hour flight from Asia to Chicago. Thankfully, he was very friendly and polite. I was searched by Mohammed because of 9/11 and the oppressive policies of Saddam Hussein. Mohammed left Iraq to escape the regime.

Next, look at circumstances where you had conflicting external and internal systems. Perhaps, your company wanted you to do something that you ethically opposed. Perhaps, you needed to take a job that was not part of your vision for your life. Perhaps, it was accepting a contract from an organization whose values you did not uphold. The internal tension came from two systems clashing.

———■———

THE KNEE BONE IS CONNECTED
TO THE SHIN BONE

Several years ago, a 70+ year-old man fell and hit his head in his garage. He was in-and-out of consciousness for a couple of weeks. As he lay dying, a parade of doctors inspected him. The lung doctor discussed how he could keep his lungs functioning. The kidney doctor suggested that he could keep the kidneys operating. Other doctors suggested this drug or that tube to keep other parts of the body functioning. The man died after several days of these discussions.

None of the doctors seemed to look at the entire system; they all focused on their subspecialty. They also seemed to ignore what the side effects of their treatment might be on the other parts of the system. None of them considered what the soul might be doing (shutting down and moving on); that was not their job. The body consists of isolated parts and if each part was optimized, the entire system would be optimized. Right? Perhaps, not quite.

Looking at parts in isolation often causes more problems than it solves.

The father of a friend of mine took a regimen of nine-or-so pills each day. These pills were prescribed by three different doctors, each one mostly unaware of what the other one was doing. All the drugs had side effects that played off each other and the old man showed disturbing new symptoms, which were ultimately traced back to the medications.

Major automobile companies used to embrace the assembly line system of manufacturing. If each person along a line did their job right, the whole car would be perfect. It often did not work that way. Now, teams work together to create parts of the car and interact to create a better product. That system creates a higher quality product than each individual working in isolation.

Reflection. Systems are not just a sum of their parts. Parts of systems interact with each other—many times, in ways we do not always understand. Looking at parts in isolation often causes more problems than it solves.

Try this. When have you tried to solve a problem only to find out that you have addressed only a small piece of a bigger puzzle? When have you fixed a problem in one part of your body only to create problems in another part?

———■———

The Columbia Disaster

In February 2003, the *Columbia* Space Shuttle blew apart upon reentry and fell to earth. All seven astronauts aboard died in the disaster. The country mourned and, as with most disasters, fingers were pointed. Who is at fault? Some months later, a 247-page report suggested that the "physical cause" of the disaster was that a piece of foam broke off during liftoff and struck the wing of the orbiter. That impact tore a hole that compromised the integrity of the *Columbia*.

But the interesting part of the report is that the piece of foam was not the only cause of the disaster. The report faults the organizational culture of NASA as much as the wayward piece of foam. In fact, the report says that NASA's "organization does not provide effective checks and balances, does not have an independent safety board, and has not demonstrated the characteristics of a learning organization" (*Columbia* Accident Investigation Board, 2003).

——■——

Most major failings result from complex systems interacting in multiple ways.

——■——

This is systems thinking in evaluating a major malfunction. In a world where individuals or corporations are blamed for extensive, complex failures, a systems approach is refreshing. What's more, it seems that the evaluators of the disaster learned the lessons from the failings of the previous space shuttle disaster, even if NASA did not. The 1986 *Challenger* disaster was also a systems failure.

Reflection. Most major failings result from complex systems interacting in multiple ways. However, litigious societies call for assessment of blame as a basis for financial remuneration. In addition, most people generally like to simplify life. The problem is that simplifying complexities leads to limited understanding. When we have limited understanding, we cannot solve problems effectively.

The *Columbia* Accident Investigation Board concluded that "only significant structural changes to NASA's organizational culture will enable it to succeed." The Board clearly rose above the event orientation to a structural orientation in evaluating the system dysfunctions that created the disaster. The Board should be applauded.

Try this. Examine a failure in an organization. How did the organization's culture facilitate the failure? What elements of the organization's culture will lead to similar failures in the future?

Then, look at a series of failures of an organization. What factors do these failures have in common? Can you trace those factors to parts of the organization's culture?

———————■———————

CLOSED OR OPEN?

In the early 1990s, a tragic episode unfolded in Waco, Texas. The occupants of the Branch Davidian compound had repulsed an FBI raid, killing and wounding many FBI agents. Days later, federal agents stormed the compound. All of the occupants ended up dead.

The Branch Davidian Church was considered a cult. One characteristic of a cult is that the members receive news from the outside world through the lens of the cult leader—in this case, David Karesh. Individual members of the organization did not have the opportunity to receive the news firsthand. Rather, Karesh fed the members news through an extremely distorted lens and they could not make accurate assessments of their environment, which led to tragedy.

Open systems thrive on rich feedback systems and input from their external environment.

The Branch Davidian Church is an example of a closed system. In closed systems, people and things on the inside do not interact with the external world. In open systems, there is much interaction between the system and the external world. Most systems fall on a continuum between totally open and totally closed. Open systems have a better chance of staying alive and vibrant in the world.

Examples of mostly closed systems abound. Other cults like Jonestown in Guyana were closed systems. The Jonestown cult ended when many of the members were coerced into drinking poison by a leader who convinced them that the external world was invading to harm them. Many family businesses that do not hire external managers have gone out of business because their systems were mostly closed. The former Soviet Union, a mostly closed system with very little interaction with the rest of the world, imploded upon itself. Abusive families stay mostly closed, fearing encroachment from the outside world.

On the other hand, ecosystems that are constantly interacting with their environments are open systems. Companies who constantly bring in new management talent and constantly solicit feedback from customers and suppliers are open systems. Open systems thrive on rich feedback systems and input from their external environment.

Reflection. Where do you see open systems around you? Where do you see closed systems around you? Can you help open up some of the systems in which you are involved?

Try this. Try employing more feedback systems within your organizations. Build mechanisms for employees and customers to have their voices heard. In one of your organizations, whether church, company, or school, notice which voices are not being heard. Help them to be heard.

———■———

BLAME ON YOU!

These stories illustrate the tendency to find a simple answer to a complex problem. Usually, we find something—or someone—to blame so that we no longer have to think about the issue.

- BHOPAL, INDIA
- JOUSTING WITH A TEAM MATE
- FIRE THE BUNCH!
- BAD BARBARA
- KICKING SAND

BHOPAL, INDIA

In 1984, a large toxic leak of the pesticide Sevin occurred outside the Indian town of Bhopal. Three-thousand residents died immediately.

Thousands more died over time from inhaling the toxic fumes. Airplanes full of lawyers flew to India to sue and counter-sue over responsibility for the deaths. Immediately after the tragedy, five Union Carbide (plant owners) managers were arrested. The president of Union Carbide, Warren Anderson, was also immediately arrested when he flew to India to see what the company could do to help the town.

As with many complex ethical situations, no one is to blame.

Let us examine some of the facts in this situation:

- The Bhopal plant was sold to an Indian operating company. Union Carbide still owned 50.9% of the plant, but the plant was run by local Indians.

- *Sevin*, a pesticide, was being produced to help the Indian agricultural system; malnutrition and starvation are problems in India—agricultural innovations were needed.

- Since taking the plant over in 1980, the Indian management cut the maintenance staff from twelve to six and reduced qualifications for skilled jobs from a college education to a high-school education.

- Manuals about safety were printed but not distributed to the population. For example: wet rags over faces could have saved many lives.

- A local ordinance stated that no one could live within 15 miles of a plant like this one. Yet, slums and shanty towns grew up close to the plant.

- False warning sirens from the plant sounded up to 20 times-a-week.

- The plant experienced high turnover and training programs were reduced.

- One employee saw the leak but said that reporting the leak was not his job.

- When the deadly cloud lifted from the plant, many people from the nearby shanty town ran toward the cloud, out of curiosity, thinking it was a fire.

- Many of the plant employees fled from the chemical cloud.

- Some employees stayed and did what they were supposed to in an emergency.

- The phones in the plant were not working, so officials could not be contacted.

Given these facts, who is to blame?

Reflection. Peter Senge (1990) suggests that in systems thinking, there is no blame, only levels of responsibility. Unfortunately, lawyers are in business to prove blame. In this case, many levels of responsibility exist from the company to the mayor, to the employees, to the management. This is a complicated situation with many actors. As with many complex ethical situations, no <u>one</u> is to blame.

Try this. The next time you are involved in a complex situation where finger pointing is the norm, stop and reflect. Map out all of the contributors to the situation. Who has contributed to this problematic situation? Extra credit: How are you contributing to a bad situation that may emerge over time?

JOUSTING WITH A TEAM MATE

Many years ago, a teammate and I worked closely in a college business program. The goal of the program was to provide major programs of study for adults who had not completed college.

Sue's job was to buy books and copies of articles for the hundreds of students who were enrolled in the program. She did her job diligently. Over time, she found that she could save the organization money by buying large quantities of books and articles at one time. She began to buy eighteen month's worth of supplies because of the price savings.

If we each did our jobs perfectly, we negatively impacted each other's performance.

My job was to develop and write high quality, up-to-date curriculum for the program. I was constantly searching out, reading, and selecting newly published books and articles so the students could have the best material for their program. When I selected the appropriate material, I gave it to Sue and she ordered the materials for the students to buy. Sue and I were partners in ensuring that students received the correct books and articles for each of their classes.

One day, I asked Sue to change the book for the Systems Management class as soon as possible. I was excited about a new book that just had come on to the market. She replied that the change would require a year-and-a-half. I was very upset at Sue for making it so hard to change books and curriculum. Sue was upset with me for being so demanding.

Mutual frustration and anger peppered the relationship.

Reflection. Sue and I played out the systems archetype *accidental adversary* (Senge, Kleiner, Roberts, Ross, & Smith, 1994) quite well. We were working together, but the system of work pitted us against each other. If we each did our jobs perfectly, we negatively impacted each other's performance. A partial solution was intense communication between Sue and myself. I began to give Sue a schedule of course revisions so she could plan accordingly.

Try this. Look at your own work situation. Is there someone who is doing his/her job well but, in doing so, making your job very difficult? When you do your best, are you wreaking havoc on someone else's work? In those cases, is it easy to place blame when you do not see the entire situation? It is easy to do your job well and also as part of your job, throw roadblocks in the way of others. This will happen naturally. It takes communication and rising above blame find a solution making place.

———■———

FIRE THE BUNCH!

The management of a large, midwestern institutional soap company perceived that they had a problem. Accounts receivable stood at 53 days; that is, it took customers an average of 53 days to pay bills. The company's financial analysts projected that the company would save millions of dollars if that payment time was cut by just a few days.

————— ■ —————

We solve problems by quickly finding leverage points of change. When these points are incorrect, people could be hurt.

————— ■ —————

Management encouraged, mandated, and cajoled the employees of the Accounts Receivable department to reduce the time from 53 days to 50 days. Nothing worked; the accounts receivable stood at 53 days. Finally, management took a drastic measure by telling the Accounts Receivable department (three people) that they would be fired if the number of day's receivable did not fall to 51 days by the end of the year.

By the end of the year, accounts receivable stood at 53 days and the entire department was fired. It turns out that this action occurred during a slight recession. Most of the customers were restaurants and were having a difficult time staying solvent. The company's salespeople told the restaurant owners and managers that if money was tight, they could defer billing. The salespeople were trying to ingratiate themselves with their customers and tighten their relationships with these restaurant owners. Therefore, the customers did not see the necessity of speeding up their billing. The casualties of this attitude were the members of the Accounts Receivable department.

Reflection. This company will never reduce accounts receivable until the sales people agree with the policy. The accounts receivable employees were used as scapegoats and fired for a problem over which they had little control. The company was obviously looking at the wrong leverage points. The real leverage lies with the sales people, not the Accounts Receivable department.

Many of us are problem solvers by nature. There is a problem—let's solve it! In fact, that is what many of us are paid to do by our organizations. Yet, we solve problems by quickly finding leverage points of change. When these leverage points are incorrect, as in the previous case, people could be hurt. When have you chosen the wrong leverage point and hurt someone or some group? When have you scapegoated someone or some group wrongly? When have you seen it happen in your organization?

Try this. Create a list of problems that have emerged in your organization or family in the past months. When you tried to solve them; what were high leverage solutions? What were low-leverage solutions? What were the differences between the high-leverage and the low-leverage solutions? Do this exercise enough and you will become a more effective problem solver.

———■———

BAD BARBARA

Barbara was technically very good at her job at a nonprofit institution in the upper Midwest. She knew her organization extremely well and performed her duties very well. She was a valued employee. Yet, she could be abrasive to fellow employees. At least one employee filed a formal complaint against her for discrimination. Her boss asked a management coach (me) to work with her to help "soften her edges"; she was perceived as the cause of her department's many organizational and communication problems.

It is always easier to have an identified patient... It is harder to take a systems perspective.

I worked with Barbara for eight-or-so sessions. I found her to be a delightful person but she could be seen as micromanaging. She worked in a male-dominated world and had to "charge ahead" to be heard or recognized. In addition, she was very direct and, as a result, could be perceived as abrasive. We worked on softening her edges and learning to be more delicate with her approaches to problems.

But something else emerged as we spoke. Many other problems also plagued the department. Her vice president was conflict-avoidant and would not deal with most behavioral problems. Other sources of friction within the department caused problems, such as some employees were allowed to come to work late; others were not. Barbara was a problem, but not the only problem.

Reflection. Barbara's temperament was a problem but not the single problem causing all of the department's problems. She is what is termed "the identified patient." All the hostility and blame in the department were focused on her. She was only part of the problem, not the whole problem.

In a global context, we make the same mistake of pointing the finger frequently. For instance, Osama bin Laden or Sadaam Hussein were considered by some responsible for all of the ills of the world. Similarly, if the United States economy hits a rough spot, we, as a nation, tend to blame the president. The truth is that reality is more complex than that.

It is always easier to have an identified patient; one can focus the blame on that person. It is harder to take a systems perspective and see how all of the parts are contributing to the problems.

Try this. Who is a problem child or identified problem in your organization or family? How are they being scapegoated for larger systemic problems? Why is it easier to blame them and not look at the bigger problems?

Look at your family, church, or organization. Who gets blamed when something goes wrong? Do you watch for that scapegoat to mess up, to reinforce your own worldview? Or do you look for the complexities of the situation?

———————■———————

KICKING SAND

I was asked to assess an organization that was experiencing significant turmoil. The Chief Financial Officer and the Chief Executive Officer were trying to fire each other through the Board of Directors. The Board of Directors asked me to investigate the situation, which was perceived to be "two boys kicking sand at each other in the sand box."

Rarely, if ever, can you point a total blaming finger at one individual.

After interviewing 21 individuals—board members, current employees, past employees— my conclusion was somewhat different. This company had systematic problems. I found that:

- The ex-CFO was fired for embezzlement.
- The ex-CEO kept a hand gun at the office.
- The current CFO and CEO had been at odds for three years.
- The current CEO and the ex-CEO had been at odds with several board members.
- One ex-employee ran up a $1000, 1-900 pornography bill; the bill had been paid off by the ex-CEO. That employee left with 1000 pornography files on his computer.
- The head of human resources had been caught riffling through trash cans after hours.
- The head of human resources was anorexic and allegedly spit her food into napkins at public lunches.
- Inappropriate comments were made by many people in the organization.
- The current CFO had made an $80 million calculation error.
- The current CEO had charged the company for a $300 golf game.

The list could go on. Clearly, this was not just a case of two boys kicking sand at each other. Many players over the years had set up a system of combativeness and fear.

Reflection. Lieutenant Calley was court marshaled for the massacre at My Lai, Vietnam. The president of Union Carbide was arrested for the chemical leak in Bhopal, India that killed 50,000 people. Some blamed the Challenger Space Shuttle disaster on cold overnight temperatures. In all of these cases, greater systematic failures were at work; they were not single point failures.

I am uneasy when clients blame their problems on single individuals or pairs of people. Usually, these "culprits" are acting out other visible and invisible forces in the system. In many cases, the identified problem is only a symptom of the other, larger problems.

Rarely, if ever, can you point a total blaming finger at one individual. If so, then another part of the system allowed his or her behavior. After observing an IT department meeting, I reprimanded an employee for his bullying behavior. But I also told other employees that they were responsible for allowing themselves to be bullied.

Try this. Look at a time in the last year when you blamed one person or organization for a problem. Then, once again, map out a different contribution of blame. What are the other pieces of the puzzle that created the situation? Can you look objectively at all the events leading to the mishap?

———————————■———————————

Fixes that Flopped

When we work in systems, our actions and interventions often have unintended effects. These stories illustrate that poorly planned solutions can lead to more difficult problems.

- Smokestacks
- Dead Cows Not Walking
- The Ethics of Birth Control
- Fixes That Fail
- Management Priorities

Smokestacks

The environmental movement burst into full bloom in the 1960s. Rachel Carson published Silent Spring in 1964, foretelling environmental disaster from chemical and other pollutants. Young people, in particular, thought that they could put a dent in this environmental degradation. They took to the streets in protest and picketed the polluters. The first Earth Day, in 1970, mobilized this movement across the country.

One of the most visible symbols of pollution was smoke pluming out of smokestacks, especially at power plants. Their visibility made them ready targets for environmentalists. Coal-fired generators produced sooty smoke that fell on nearby neighborhoods. Grandma's clean white sheets on outdoor clotheslines turned gray.

Good intentions are only that— good intentions. Moral high ground becomes low if the wide angle of systems are not considered.

Environmentalists and others called for higher smokestacks to launch the soot higher in the air to be carried away. The smoke contained very acidic sulfur dioxide and nitrogen oxides. Electric utilities built higher smokestacks and the nearby neighborhoods were partially spared.

Many of these coal-fired power plants were located in the Ohio Valley. The smoke was carried farther northeast into the hardwood forests of New York and Canada. The soot fell out and acidified lakes and woods (termed acid rain). This terrain did not contain the alkalines that the Ohio Valley has to naturally buffer itself from the acid. Fauna and flora were (and continue to be) poisoned.

Furthermore, due to pressure from environmentalists, electrostatic scrubbers were installed in these smokestacks to remove the soot. This technology was successful. However, the soot partially neutralizes the sulfur dioxide and nitrogen oxide when it lands. The soot, ironically, shields the earth from acid rain. Without the soot, this acid rain became much more deadly.

Reflection. Good intentions are only that—good intentions. Moral high ground becomes low if the wide angle of systems are not considered. I was one of those environmentalists. I perceived a problem and wanted a quick solution. The systems principle, *the easiest way out always leads back in,* (Senge, 1990) is true in this situation. Our quick, knee-jerk solutions create additional problems. This is also an excellent example of the principle *there is no away.*

Try this. Take a personal inventory. When have your good intentions created poor results? Could you use this historical situation to anticipate other problematic situations in the future?

Source: Tenner, E. (1996). *Why things bite back: Technology and the revenge of unintended consequences.* New York: Vintage Books.

———————■———————

DEAD COWS NOT WALKING

Several years ago, the mad cow disease scare spread through the United States. Outbreaks of mad cow disease in Europe and Canada made people fearful about acquiring the disease from eating and dying from infected beef. In response, the United States Department of Agriculture "swiftly imposed new meat handling rules." Meat packing plants cannot process crippled cows because they might have the dreaded mad cow disease. Most, if not all, of these cattle are crippled but do not have mad cows disease.

Quick political fixes often create unintentional consequences which breed work-arounds that are not necessarily productive.

Therefore, farmers are left with a dilemma. They can butcher the crippled cows themselves under less sanitary conditions than the plants or burn or bury the bodies, thus incurring a financial loss. They can ship the cattle to a black-market processor who will butcher the beasts under the radar of the USDA. In all these cases, however, the cattle cannot be tested for the disease. Valuable data about the spread of the disease is lost and the USDA's efforts are thwarted.

This is another example of the "unintentional consequences" of a quick fix. The USDA was pressed to restore public confidence in eating beef. It created a dramatic policy that was virtually unworkable. Cattle can only be tested if they are brought in, but if they are brought in, the farmer loses his investment.

Most, if not all the crippled cattle are only crippled, not diseased. In fact, many become crippled in cramped trucks on the way to the processing plants. But, if the cattle arrive crippled, they must be destroyed. On another level, this is a societal waste.

Reflection. The government, national or state, due to political pressures, often feels they must impose a quick fix—politically visible solutions to complex problems. As a result, unintentional consequences often occur. In this case, new knee-jerk regulations push an industry to expend their energy creating workarounds and ways to beat a system.

Try this. When have you found workarounds to new policies? When have you observed others finding loopholes to beat inefficient policies? Quick political fixes often create unintentional consequences which breed workarounds that are not necessarily productive. When have your workarounds hurt other systems or other people's efforts?

THE ETHICS OF BIRTH CONTROL

In the 1960s, the Chinese government realized that it had a major problem: too many people. The history of China is punctuated by famines. Therefore, the Chinese government decreed a new policy of one child-per-family. Families not complying would be subject to fines and heavy taxation.

We cannot always predict the consequences of large actions. We can only look at possibilities.

Yet, besides the government system, other systems were at play. The family farm system valued boys above girls because boys are economic entities and can run the family farms better. Girls required dowries. So, if you only get one child, better make it a boy.

And so, over the past 40 years, some Chinese have taken steps to ensure that their only child is a boy. There being reports of infant girls have been killed at birth and buried in the countryside. In some cases, ultrasound machines (invented to detect deformed babies) have been used to detect fetal gender and, if female, an abortion is performed. Some baby girls have been left at an orphanage or police station.

Over the course of years, the unintended consequence of this governmental policy has been to create a population imbalance. In China 2005, there were 120 men to every 100 women. The government is sending its young men to the United States and elsewhere for a college education (and hopefully, a bride). Young women from the Philippines and elsewhere are being imported as brides.

But, another unintentional consequence is that this Chinese system rubs against another system. Birth rates are falling in the United States due to later marriages, stressful careers for men and women, overt gay lifestyles, and other factors. Many couples, especially older couples, are adopting babies from China. I was on a plane going from Hong Kong to Chicago with 13 newly adopted, ten-month-old girls (yes, all girls) with their delighted and elated adopting parents.

Reflection. A governmental problem and a policy aimed at a solution have huge unanticipated consequences with worldwide ramifications. I am not sure if the authors of this policy had these outcomes—some positive, some negative—in mind when drafting it. We cannot always predict the consequences of large actions. We can only look at possibilities.

Try this. Look at some recent actions taken by your local, state, and federal governments. Can you predict any unintentional consequences of these actions or new laws? Sometimes, you can also look at actions not taken to see the unintentional consequences. Proposals by the Army Corps of Engineers to strengthen the New Orleans levees were turned down for years and the result was the 2005 devastating floods in the wake of Hurricane Katrina.

FIXES THAT FAIL

- Many years ago, I was a teacher's aid at a Catholic elementary school in Jersey City, New Jersey. One child had a particularly disruptive morning. The classroom teacher made him stay inside during lunch and recess. Guess what his behavior was like during the afternoon? Not able to use his pent up energy during recess, he was a terror in the afternoon!

- A small, midwestern college had three different registrars over a short period of time. After three in a row failed, it was time to get a new one. Each of the previous three had failed for different reasons. One registrar did not understand the college's enrollment systems; another one did not have the right disposition; and so on. Clearly, the job was untenable for any registrar. Unless the nature of the job changed, the next registrar would also fail.

The problem is that many problems lead back to their original causes.

- When my children were young, I would ask them to wash the dishes after a meal. They usually did a poor job and, maybe, only a third of the dishes would get done. I did the dishes. As my father said, "If you want something done right, you need to do it yourself."

In all these cases, the quick fixes failed. Keeping the spirited child inside during recess only made him more unruly during the afternoon. The registrars were not bad people; the systems around the position did not support them. Doing the dishes for my children only increased their dependence on me and did not help them learn how to do the dishes any better.

Reflection. We are a problem-solving oriented people. We love to solve problems as fast as possible. Some of us are even paid to be problem solvers. The problem is that many problems lead back to their original causes. We call these *fixes that fail* (Senge, 1990).

Try this. Look at some fixes that you or your organization has implemented in the past year. When have those fixes backfired? What have been the consequences of those failures? Could you have seen, in advance, the likelihood of the fix failing? What would have been your clues?

———■———

MANAGEMENT PRIORITIES

During the 1970s, a midwestern industrial chemical company had a 700-member sales force. Sadly, over 300 of these jobs turned over each year. This turnover created a huge, direct expense in recruiting, selecting, and training and also the indirect expense of discontinuities in customer service.

The District Manager had incentive for the new employee to fail and little incentive for him to succeed.

Management blamed the full-commission compensation system for the employee turnover. New sales people got paid only after they made a sale. In industrial sales, however, it often took five or six months to make a sale. The sales process might include product trials, introductory meetings, trust building, and other steps. Therefore, unless the sales person received a salary draw against their future commission, they might go months without any income. That compensation system drove new sales people from the company after a few months; they could not support themselves financially.

The company's solution was to only hire young men (Remember, this was the 70s.), who lived with their parents, as sales people. These men could live at home while they built their sales and commissions. However, that solution did not seem to help. Young sales people were just as easily discouraged by lack of immediate sales as previous employees.

The managerial compensation system created this retention problem. The majority of the District Manager's compensation came from the commission on *his* own accounts. Only a small amount of his compensation was pegged to a new sales person's success. After training a new employee for a week, the District Manager would send him out to build his own sales. The new employee, poorly trained, often floundered and drowned. His success or failure had little impact on the District Manager's personal bottom line.

As an added kick, when a new employee failed, the District Manager could pick up his good accounts and keep them as the manager's own. The District Manager had incentive for the new employee to fail and little incentive for him to succeed.

Reflection. Compensation systems are one of the first places to look if things are amiss in organizations. Often the compensation system sends employees' behavior in a direction other than the organization's stated goals. The unintended consequences of compensation structures often collide with the company's goals.

Try this. Look at the reward systems in your work place and in your home. Are you rewarding the desirable behaviors or do loopholes support undesirable behaviors?

————————■————————

Pushback, Blowback, and Unintended Consequences

These stories illustrate that when you push the system, the system pushes back. Systems often resist our intervention (pushback), create situations that are the opposite of what was intended (blowback), or create a myriad of new problems (unintended consequences).

- Plumbing, Anyone?
- Bob's Hip
- Would You Like Fries with That?
- Grilled Salmon Steaks
- Mobile Meth Labs
- Tuvalu, Honey
- Compensating Feedback

PLUMBING ANYONE?

In North African villages, women gather by the wells in the morning and wait for their turn to draw water. On their heads, they balance huge jars used to carry water to their homes. Because the wells are deep and small, the women often have to wait for several hours. They chat in line, talking about their lives.

Examine how social and technological systems interact and collide in your workplace when new technology is introduced.

The United States Peace Corps witnessed this and wanted to help. They saw how the women lost hours of time each week, waiting for their turn in line. So, the Peace Corps built plumbing systems to deliver water directly into the homes of these women—no more walking, waiting, or wasted time.

The Peace Corps built the plumbing systems, constructed the pipelines, and brought water into the homes of these North African women. They were proud of their work and believed that it was truly helping these third-world women. However, the plumbing was broken within a month. The Peace Corps volunteers fixed the plumbing but it broke again. They fixed it. It broke again. What happened?

Reflection. Two systems collided here—the social system and the technological system. In the context of United States culture, which reveres productivity and sanitation, the Peace Corps solution was logical enough—indoor plumbing meant no more wasted time at the village well. But for these women, the time at the well was their only social time of the day, the only time that they were allowed outside their houses. This was important social time, not wasted time. So, the women sabotaged the plumbing.

Is this an isolated example relegated to North Africa? No. Examine how social and technological systems interact and collide in your workplace when new technology is introduced; it upsets the social system and pushback results.

Recently, one of my clients decided to allow clients to complete transactions through the World Wide Web. The members of the IT department, who were in charge of the Web, were elevated in status and importance in the organization. Those employees who controlled access to other applications lost importance. Bad feelings ensued.

Try this. Examine a situation in your life where social systems and technological systems collide. Perhaps, your children are having Instant Messaging conversations with their teenage friends while you are trying to have a family dinner. Perhaps, collision happens when a new technology is introduced in your work place. What are the results? How can the transition be smoother?

———————■———————

Bob's Hip

Bob, the dean of a midwestern college, is a nice guy, who does not deserve to have anything bad happen to him. Yet at the age of 53, he had a hip replaced that had been steadily deteriorating and causing him great pain.

Cause and effect are not related in time and space.

What had he been doing recently to cause this problem? Nothing. He played a little basketball, loved golf, and took walks with his wife. He was not doing anything that a 53 year-old man should not do. What happened?

Between the ages of 14 and 24, Bob was a baseball catcher. Catchers crouch low, receive a pitched ball, stand up, and throw the ball back to the pitcher. In the process of throwing the ball back to the pitcher, a catcher pivots on one hip as he stands up. How many times did Bob perform this motion in those ten years? We do not know. His doctor suggested that the repetitive motion process wore out his hip joint.

When Bob decided to be a catcher, little did he imagine hip-replacement surgery would be a result. Would his life and health been different if he would have chosen to be a left fielder? We will never know.

Reflection. Cause and effect are not related in time and space (Senge, 1990). Bob stopped being a catcher almost 25 years before his problems started. Yet, he had worn out his hip. How many other baby boomers, caught up in the fitness generation, have compromised parts of their anatomy in the name of fitness? Long-term consequences of repetitive actions are often difficult to gauge. I would invest in artificial knees and hips as baby boomers continue to age.

Try this. Notice all of your movements and activities (or non-activities) over the course of a month. Is anything that you are doing setting you up for problems later? Is red meat slowly clogging your arteries? Will playing copious video games cause ocular lock or other visual problems? Is your fast-paced life setting you up for high blood pressure problems? Is listening to heavy rock music creating a hearing problem? Is using coffee, alcohol, cigarettes, or chocolate creating an addiction problem that could be hard to kick? (I am suddenly worried about all of the typing that I do. Am I setting myself up for repetitive stress syndrome? Am I not flossing enough? I need to read my own book.)

————■————

Would You Like Fries with That?

Ray Kroc was probably not thinking about his own or anyone else's gall bladder when he eyed the little milkshake stand owned by the McDonald brothers in 1954. He was reported to be thinking about the efficiency of several milkshake machines as they whirled around making milkshakes for hot California motorists on that summer day.

Often cumulative effects are the effect of long-term causes and not an immediate cause.

He bought out the McDonald brothers and expanded the restaurant chain. McDonalds and other fast food chains became essentials for a mobile American population. Life in a car in the fast lane necessitated different eating habits. The landscape became dotted with fast food restaurants that catered to that lifestyle over the next 50 years. Ray Kroc died a multimillionaire. Hamburgers, french fries and a milkshake became standard American fare.

But a funny thing happened on the way to this food nirvana. By the year 2000, gall bladder surgery became the most common surgery in the United States. Gall bladder problems are most commonly linked to eating fatty and greasy foods. The rise of eating fast food meals seems to coincide with the rise of gall bladder problems, though Ray Kroc could have never predicted it.

In addition, obesity has reached epidemic proportions in the United States. This problem is also partially the result of people eating fast foods on the run.

Reflection. Cause and effect are not closely related in time and space is a systems truth, according to Peter Senge (1990) and others. If an individual eats a burger and fries today, the gall bladder will not explode tonight as a result. Cause and effect are often separated in time by lag variables. Often, long-term processes create cumulative effects. Ray Kroc, unknowingly, helped to create an industry of a special type of surgery.

Try this. Look around your home and life for time savings and useful inventions. What unintended consequences come from those inventions? Do video games create sedentary children? Does the channel changer help create a passive lifestyle? Does air conditioning create an indoor lifestyle that results in our not meeting our neighbors? How can we overcome these unintended consequences?

———————■———————

GRILLED SALMON STEAKS

I love grilled salmon steaks, especially when marinated with a little orange or lemon. I especially like it when the salmon is only $3.99 per pound. This is an affordable luxury. Or is it?

The only salmon that is inexpensive is farm-raised. Problems exist with farm-raised salmon because they are kept in "ocean feedlots." These feedlots create uncontained pollution—"200,000 salmon generate as much fecal material as a city of 60,000 people." Some contained salmon have higher PCB levels and certain pesticide levels than are allowed by EPA standards. Salmon are known to escape their pens, through flooding and other natural occurrences. These escapees mate with wild salmon, compete with wild salmon, and contaminate wild salmon with sea lice and other illnesses. Farm-raised salmon depend upon sardines, anchovies, and herrings for food. Some estimates suggest that it takes 5 to 6 pounds of ground smaller fish to produce a pound of salmon.

There we have it. My "inexpensive" salmon do not taste better, are often less healthful for me, could create hazards for natural salmon, and hurt the environment. Should it really cost only $3.99 per pound? The real cost to our society is much more.

My family once bought my young son two gerbils for $3.99 each. Even with the approximately $50 worth of tubes and other cage apparatus, the gerbil project seemed like a bargain for teaching him about animal psychology, responsibility, and caring for others. However, gerbils do get away; they chew on clothes, electrical cords, telephone cords, and other valuable items. The gerbils ended up costing us hundreds of dollars. (Thankfully, the worst offender ended up trying to swim in the toilet and drowned.)

Reflection. It is difficult to assess the real cost of goods. Government subsidies and unanticipated consequences of production mask the actual cost of items. Automobile users benefit from highway subsidies and the exhaust from vehicles create a greenhouse effect. It is argued that smokers hurt the lungs of others with secondary smoke. Pork feedlots create environmental cesspools.

Try this. Think of some consumable that you use. Can you trace the total costs of producing and distributing that product? Who, if anyone, was harmed in the production and distribution of that product?

In other words, what is the real cost of that shirt that you are wearing? What is the real cost when you account for the 12 year-old Malaysian girls who sew them for $1.25 per day, 12 hours-a-day, with only one bathroom break—cheap but costly.

———————————■———————————

MOBILE METH LABS

We hear it constantly in the news. Methamphetamines are ruining the lives of our young people, especially in rural America. Meth is addictive and turns users crazy. Some users commit suicide while on it. It is easy to make. If you have a kitchen, you can create a manufacturing lab because the ingredients are found in leading cold medicines. Distill them; add a little paint thinner and you have meth.

The federal government took action by providing block grants to states for better enforcement. One state used the grants, put excellent enforcement in place, and ran the labs and users out of the northern part of its state. Guess what happened in the southern part of the state immediately to its north? You bet. An infestation of labs and users showed up.

There is no away. Problems move but do not disappear.

Some states took other actions. They made the cold medicines containing active meth ingredients available only with a prescription rather than over-the-counter. Guess what happened? The users and labs went into nearby states to purchase the cold medicines over-the-counter. Of course, as an unintended consequence, the more rigorous states lost sales tax revenues.

Some stores took action by limiting the number of cold medicine boxes that individuals could buy to three. But users and producers went to other stores that would sell them more. Of course, the more rigorous stores lost sales revenues. Some sales clerks advised customers to come back and buy three boxes as often as they wanted. The stores rigor became an inconvenience rather than a real deterrent.

Meanwhile, despite great intentions, meth spreads. Labs can be created and taken down literally in minutes; producers know no state, city or store boundaries.

Reflection. There is no away. Problems move but do not disappear. People involved in meth or other drugs do not quit because one state or store makes it more difficult to obtain the necessary ingredients. Vice finds a way of beating haphazard impediments.

In addition, good intentions often do not lead to good outcomes. Often, isolated good intentions hurt the intender.

Try this. There is no away. How does this work in your life? What do you want to go away that will not go away? Does that problem help you dig deeper into your repertoire and stretch yourself and, thus, help you grow? Think about it.

———————■———————

TUVALU, HONEY

I drive to the coffee shop and enjoy a cup of coffee while working in air conditioned comfort. I use my laptop, listen to the espresso machine, and watch cars drive by. Later in the day, I will drive my son to a friend's house, find some fresh food (shipped from somewhere else) in my refrigerator, make a lunch, and maybe watch a movie. It is a typical day of an American citizen.

The average Tuvaluan of my gender and age awakens to a beautiful sunrise over the Pacific Ocean. He probably takes his small fishing boat out to sea and throws his nets into the water. His wealth that day will be determined by how many fish he pulls up in his nets. It is a simple life compared to mine. Why should he be concerned about my life and what I do?

We all affect each other in ways that we hardly realize and, sometimes, it is not pretty.

He lives in Tuvalu, the fourth smallest country in the world, an archipelago consisting of nine major atolls scattered in the Pacific Ocean located north of New Zealand and the Fiji Islands. To a westerner, this country seems like a cross between Gilligan's Island and heaven. A beautiful, blue sky falls into the silvery blue waves as fish and dolphins play in the surf. But the highest point of the islands is sixteen feet above sea level and the islands are sinking.

The islands are sinking because of my use of electricity and gasoline in North America. Granted it isn't all just me but also millions of Americans, driving SUVs to work and school, using computers, air conditioners, lights and other trappings of modern technological civilization. The use of fossil fuels is apparently heating up the atmosphere which is melting glacial icecaps and raising the ocean levels. It is predicted that this small country will disappear because of global warming; it is the "proverbial canary" in the poisoned coal mine.

As one resident of Tuvalu says, "This is the sharp edge of the climate-change debate... Forget politicians, scientists, and activists. What it boils down to are waves in my bedroom."

Reflection. Why should I worry about a bunch of people in a country that I have never heard about? Because my actions impact their lives, livelihoods, and ways of being. Why should they worry about me? Because I am taking away their way of life and their ancestral homelands. We all affect each other in ways that we hardly realize and, sometimes, it is not pretty.

Try this. Consider one of your actions. What ripple effect does it have within your community or in distant communities? Map out the relationships and effects of your "cause."

Source: Levine, M. (2002, December). Tuvalu, Toodle-oo. *Outside Magazine*

COMPENSATING FEEDBACK

One systems truth is *the harder you push, the harder the system pushes back* (Senge, 1990). Every time we push when there is a system in place, we can expect the system to push back in some way. Let's see how this works.

- In the 1960s, activists, mainly from the northern part of the United States, built voter registration drives for African-Americans in the South. The social and political system in the South mostly excluded that part of the population. That system pushed back by resisting voting rights and, ultimately, in the murder of civil rights workers.

Every time we push when there is a system in place, we can expect the system to push back insome way.

- My alma mater college became co-ed in the early 1970s. Many students and alumni were disappointed with that decision. The existing system was an all-male social and educational system. Pushing women into that system created a backlash. Women were ridiculed, slurred, harassed, and treated poorly by a few men at the beginning of that change.

- A union goes on strike because management tried to rollback their wages and decrease their privileges. Management pushing on the existing system created a pushback from the union.

- I get up in the morning and realize that I need to go on a diet—my middle-age bulge is expanding. I forego breakfast and lunch. Perhaps, I drink juice or tea. I am pretty proud of myself. At about three o'clock in the afternoon, I walk by a colleague's desk that has a bowl of dark chocolates and I eat it, without restraint. I am pushing at my body system and it pushes back with vengeance.

In all of these cases and many others, existing systems are pushed and, then, push back.

Reflection. Where do you see compensating feedback in your life, workplace, or family? The next time you feel that you are on the receiving end of someone else's major change, breathe deeply and observe your own feelings and thoughts. Are you reacting viscerally, immediately? Do you react even before you rationally weigh the costs and benefits of the change? Feel the compensating feedback within you.

Try this. Impose a small change on people around you and watch their reaction. Go to work with orange hair, make your children eat broccoli, or don't take out the garbage. The reaction you get is compensating feedback. The stronger the feedback, the stronger the system you are pushing against.

———■———

HIGH-LEVEL INTERVENTIONS

In systems, the consequence often comes from an unanticipated intervention. These stories illustrate our initial reaction is not always the most effective. Often, our most powerful interventions are counter-intuitive.

- HUSH PUPPIES
- SLEEP IN THE OUTHOUSE
- PUT IT ALL TOGETHER
- CARRYING CAPACITY

HUSH PUPPIES

Remember those soft Hush Puppy shoes that were popular in the 1960s? They almost died a sad, neglected death. But now they are back.

By 1994, only 30,000 pairs of Hush Puppies were sold. The line was almost dropped by Hush Puppies because of lackluster sales. But 430,000 pairs were sold in the following year and that number quadrupled the following year. Did Hush Puppies create a brilliant marketing plan or execute an incredibly effective media blitz? Not really. No one was trying to make Hush Puppies a trend.

Leverage points come from unexpected places.

A few hip kids began buying new and used Hush Puppies in odd, little stores in lower Manhattan. They wore them to hip bars and nightclubs. This trend was picked up by two fashion designers, who featured Hush Puppies in their spring fashion lines. Then, a west coast designer picked up this line of shoes. By 1996, the Council of Fashion Designers gave Hush Puppies a prize for the best accessory. Sales were booming—up to a half-million pairs a year.

Reflection. Leverage points for real change come from unexpected places. Often, they cannot be planned but are only reacted to. When you have leverage, little things make a big difference. No one planned or executed this epidemic in Hush Puppies. They simply became "cool" again. This coolness became infectious; sales spread like an epidemic. What other products could be retrieved from mothballs and developed into epidemics?

Try this. Look at the history of an organization close to you. Where have small changes created major results? Where are the tipping points where small changes make big results?

Source: Gladwell, M. (2000). *The tipping point: How little things can make a big difference*. Boston, MA: Little, Brown, and Company.

———■———

SLEEP IN THE OUTHOUSE

I had the privilege of being a "dorm dad" for 15 inner-city, fourth and fifth grade boys at the environmental camp Widgiwagon in northern Minnesota. Many of these children had never been away from their parents or out of the Twin Cities, let alone the wild woods of the far north. My responsibilities included preventing the cabin from burning down, not allowing anyone to drown or die, and, generally, keeping the peace.

High leverage actions often come from unexpected places.

The first day of this five-day adventure included a long bus ride, orientation of the facility, and a night adventure. After that, the dorm dad was tired, but the boys were wired. So, I allowed some ruckus between nine and ten o'clock and then turned the lights out and demanded complete silence. The boys needed their rest before the long day ahead.

It did not quite work that way. Slumber parties often do not involve sleep. The wise, college-professor dorm dad attempted a series of the low-leverage, useless interventions:

- "Shut up!! Or you will get in big trouble!" (threat)

- "Please go to sleep. You need your rest." (projection)

- "Please be considerate of others." (reason)

- "Shut up or the bear will hear you!" (fear)

- "The next one who talks will sleep in the outhouse!!" (ridiculously stupid, desperate threat)

None of these approaches put a dent on the noise level. It was 11:00 p.m. and the author/dorm dad/disciplinarian/systems thinker was tired, grouchy, and irritable, and out of threats.

Out of the blue, I said softly, "Dear God, thank you for bringing Cardell on this trip. He is so nice to his classmates and tries so hard to be a good student."

Pure silence.

"Dear God, thank you for Joey. He is so happy that he cheers up the other kids. He also helps them remember how to behave some times..."

Pure silence.

The boys hung on every word that I spoke, waiting patiently to hear what I would say about each and every boy. No more talk; everyone went to sleep. Some of them had never been prayed for before. For some of them, it was rare for someone to say something nice about them. I did the same thing every night for the next several nights with the same result.

(After the first night, my son approached me and said "Dad, you said that Adam was a sweet boy. Dad, he isn't.")

Reflection. High-leverage actions often come from unexpected places. This high-leverage action was not pre-meditated but spontaneous, born out of desperation. Often, we have to bang our heads on low-leverage interventions before a high-leverage one pops up. In all fairness, the teachers in the girls' cabin could not pray because they were public school teachers and there is division of church and state. But what could they do to me, fire me? Or, I suppose, I could have slept in the outhouse....

Try this. Be more conscious about a solution that is not working. If your tried-and-true methods are not solving a problem, try something radically different. What do you have to lose?

PUT IT ALL TOGETHER

In the mid-1990s, I was part of an organizational-development intervention at a large financial service company. We brought the entire Human Resource department together for two days in a large hotel banquet room. All the members of the benefits administration, compensation, training, payroll, and other sub-departments attended. Even though these individuals were in the same department, many of them had never been together in the same room.

Whatever impacts one part of a system impacts the other parts.

During the two days, we addressed many common problems of the entire department and attempted to take a system-wide approach, rather than isolated approaches, to these problems. At one point, members of each subunit sent the other subunits flipchart paper with lists of "what we appreciate," and "what we need more of" so they could better understand each other. Positive results emerged from this intervention. Better communication and working relationships were created.

Our secret: we had brought the entire system into the room to find solutions. A system principle suggests that *you can't divide an elephant in half.* That is, you cannot understand an entire system by cutting into pieces. Whatever impacts one part of a system impacts the other parts. Changes create a domino effect. Therefore, trying to change one piece of a system without addressing the other parts, will achieve limited success.

These kinds of large-scale interventions were developed by organizational consultant Kathy Dannemiller. She held a large system intervention with 7000 workers from Boeing, the aircraft manufacturer. The result was that Boeing dramatically reduced the cycle time for developing a new airplane. Another whole-system intervention brought together members of the US Forest Service, environmental groups, the Environmental Protection Agency and forest companies, such as Weyerhauser and Potlash. The focus of the intervention was how to manage our country's forest lands. Again, all the parts of the system concerning the forests were brought together and communication between all the elements helped the process.

Reflection. Have you tried to solve problems when all the parts of the system were not present? What success did you have? Compare these efforts to when you had all the pieces of the system together and attempted problem solving. Which worked best?

Try this. Next time you have a major problem to solve, bring all parts of the system, or as many as you can, together. It may feel awkward; people may wonder what they are doing there, but there could be some surprising results.

———————■———————

CARRYING CAPACITY

Many famines have plagued northern Africa in recent years: for example, Somalia, Ethiopia, and Chad. The Sahara desert has expanded, eating away at tillable land. Due to better medical practices, populations have grown. The result is famine. The population outstripped the carrying capacity of the land.

In systems thinking parlance, this is a *limits to growth* archetype (Kim, 1994). The growth of one system cannot be sustained in the context of another. For instance, if the Minnesota deer population is allowed to grow unchecked by natural predators, there will be a die-off, because the land can sustain only so many deer, given the amount of food and space. Unchecked population growths will approach limits that cannot be sustained.

This principle is also true in commercial systems. In the late 1990s, AOL, the Internet provider, crashed for several days. Because they had added thousands of new subscribers, their entire technological system crashed. Their systems could not sustain the growth of new clients.

Conversely, Simon Delivers, a grocery-delivery business based in Minnesota, stopped taking new clients for several months in the late 1990s because it did not have enough warehouse and truck capacity for new clients. Contrary to conventional beliefs, stopping an influx of new customers was not the death knell of Simon Delivers, but rather the management decision that sustained it. Many other grocery-delivery services have folded; Simon Delivers thrives. Simon Delivers built a strong infrastructure in several months that allowed for future growth.

In the 1980s, a specialty chemical company limited the bonus and commission received by its sales force to the first 18% of growth over sales of the previous year. In other words, it discouraged growth over 18% per year. The company did not want its sales force to sell more than it could effectively service. The company was the most successful company in its industry.

Reflection. Biological and social systems have a limited carrying capacity. After that point, growth is limited by these forces. On the other hand, companies and governments can be aware of these natural barriers and plan growth accordingly.

Try this. What is your carrying capacity? Your family's? Recent studies report that our children are overscheduled, being shuttled between piano lessons, soccer practices, and school (without a family dinner). Kids don't complain, but what is their carrying capacity? Adults often collapse into illness before they realize that they have reached their carrying capacity. What about you?

———■———

CONCLUSION

In 2005, the United States and particularly New Orleans was hit hard by the effects of hurricane Katrina. That disaster is full of lessons about systems thinking. New Orleans was founded below sea level over 200 years ago—cause and effect are often not connected in time or space. The Army Corps of Engineers was only given limited funds to make New Orleans safe—quick fixes often boomerang back on us. Most of FEMA's funding is going to anti-terrorist measures, not natural disaster prevention or relief- you can have your cake and eat it too, but not all at once. Disaster relief organizations did not act in concert with each other—whole systems need to solve tough problems. One could even write an entire systems book on the New Orleans systems failures.

So, do you really want to be a systems thinker? On the downside, you could end up irritating people. When a group is ready to find a solution to the problem, you might ask "What are the unintended consequences of that action?" Or "The easiest way out will lead back in." You will eschew the world of simple solutions. Problems that seemed simple are no longer simple and could become frustrating. Finally, you will have yet another discipline in your life.

On the upside, you will be a better thinker and person. You will be a stronger problem solver. You will be a more marketable employee. You will have more wisdom to offer others. You will not ask why simple solutions do not work. You will not ask dumb questions. You will engage in less judgment, blaming, and scapegoating; the world will be better for that.

Ultimately, you can write some of your own systems stories. Some of them will undoubtedly be better than the ones that you just read. Your life, in fact, could probably fill a book like this of good, insightful stories. Now that you are thinking like that, keep it up—the world will be a better place.

BIBLIOGRAPHY OF BOOKS
MENTIONED OR CITED

Bowen, M. (1985). *Family therapy in clinical practice*. Lanham, MD: Jason Aronson Publishers.

Carlson, R. (1964/2004). *Silent spring*. Boston, MA: Houghton-Mifflin.

Columbia Accident Investigation Board. (2003). *Report of the Columbia Accident Investigation Board: Final Report*. Available at http://caib.nasa.gov/default.html

Gillon, S. (2000). *That's not what we meant to do: Reform and its unintended consequences in the twentieth century*. New York: W.W. Norton and Company.

Gladwell, M. (2000). *The tipping point: How little things can make a big difference*. Boston, MA: Little, Brown, and Company.

Haines, S. (2000). *The complete guide to systems thinking and learning*. Amherst, MA: HRD Press.

Hardin, G. (1968). *The tragedy of the commons*. Science, 162, 1243-1248.

Kauffman, D. (1980). *Systems one: An introduction to systems thinking*. Minneapolis, MN: Future Systems.

Kim, D. (1999). *Introduction to systems thinking*. Waltham, MA: Pegasus Communications.

Kim, D. (1994). *Systems archetypes 1: Diagnosing systemic issues and designing high-leverage interventions*. Waltham, MA: Pegasus Communications.

Meadows, D. (1972). *Limits to growth*. Colchester, England: Signet Publishers.

Senge P. (1990). *The fifth discipline: The art and practice of the learning organization*. New York: Doubleday Currency.

Senge, P., Kleiner, A., Roberts, C., Ross, R., & Smith, B. (1994). *The fifth discipline fieldbook: Strategies and tools for building a learning organization*. New York: Doubleday Currency.

Tenner, E. (1996). *Why things bite back: Technology and the revenge of unintended consequences*. New York: Vintage Books.

INDEX

About the Author

Dr. Richard Brynteson, author, consultant, and speaker, is an Associate Professor of Organizational Management at Concordia University in St. Paul, Minnesota. During his 25 year career, he has worked and lived in South America, Asia, the Middle East, and Europe. His current occupational mission is to push the systems, creative, innovative, ethical, strategic, futuristic, critical, and emotional thinking of his students and his clients. He has two beautiful teen age children who constantly teach him much. Richard has a B.A. in English from Dartmouth College, an M.B.A. in Marketing from the University of Chicago, and a PhD. in Education from the University of Minnesota.

Richard can be reached at Brynteson@csp.edu .